# 1906 Earthquake

## San Francisco's Great Disaster

Sandor Demlinger

Schiffer Publishing Ltd
4880 Lower Valley Road, Atglen, PA 19310 USA

I dedicate this book to

my wife Reva,

my sons Keith and Glenn,

and my daughter Allyson.

Published by Schiffer Publishing Ltd.
4880 Lower Valley Road
Atglen, PA 19310
Phone: (610) 593-1777; Fax: (610) 593-2002
E-mail: Info@schifferbooks.com

For the largest selection of fine reference books on this and related subjects, please visit our web site at **www.schifferbooks.com**
We are always looking for people to write books on new and related subjects. If you have an idea for a book please contact us at the above address.

This book may be purchased from the publisher.
Include $3.95 for shipping.
Please try your bookstore first.
You may write for a free catalog.

In Europe, Schiffer books are distributed by
Bushwood Books
6 Marksbury Ave.
Kew Gardens
Surrey TW9 4JF England
Phone: 44 (0) 20 8392-8585; Fax: 44 (0) 20 8392-9876
E-mail: info@bushwoodbooks.co.uk
Website: www.bushwoodbooks.co.uk
Free postage in the U.K., Europe; air mail at cost.


Library of Congress Control Number: 2005938681



Designed by Mark David Bowyer
Type set in GoudyHandtooled BT / Zurich BT

ISBN: 0-7643-2404-7
Printed in China

# Contents

**Prelude to Disaster:** San Francisco Before the Quake .................... 4

**Giant Waves & Falling Buildings:** The Earthquake and Fire .......... 68

**Rebuilding Hope:** The Aftermath ................................................ 119

# Prelude to Disaster
# San Francisco Before the Quake

The warning was clear as a fire engine's bell. In May 1893 *The City Argus* printed a story about the dangerous condition of San Francisco's fire fighting equipment. Fire Chief Sullivan asked the city for 2,000 more fire hydrants, ten engines, five chemical and five hook-and-ladder companies that year. Various sections of the city had no hydrants and no fire fighting equipment whatsoever. Moreover San Francisco was isolated and there was really no adjoining town to call for help. Oakland, across the bay, had only two fire engines that could be spared. To make this point clear, the paper cited a recent fire in New York City that required some twenty-five engines to put out the fire. "Don't wait until a big fire breaks out to find out why it spreads. It is up to the supervisors of the city to make the appropriation in June. If nothing is done about this problem, you cannot blame Chief Sullivan." *The City Argus* was calling for help.

It was noted in the same issue that the San Francisco supervisors were growing tired of paying the water company $91.50 for every new fire hydrant and $30.00 for its use every year. In the original contract, for the privilege of laying mains in the streets and alleys and collecting money from the consumers, the water company agreed to supply the city with water free of charge for extinguishing fires, sprinkling streets, and irrigating the public squares. This did not happen.

Other news from the 1893 *City Argus* sounds a lot like today's news. Corruption in the city. The town was being overrun with bogus lottery tickets. Douglas, the custom house smuggler, procured bail. Would the Grand Jury ever find the election fraud perpetrators? After May 5th, all Chinese not registered according to the Geary Act were to be deported. Why were the electric-car employees not being paid? This was not news. San Francisco was always cleaning up its act, and as far back as 1856 the Committee of Vigilance sought to clean up the city and improve the business prospects.

Before gold was discovered there in 1848, San Francisco was a quiet frontier town. On July 9, 1846, Captain John Montgomery came ashore at Yerba Buena with 70 troops and in the center of town he read a proclamation claiming the territory for the United States. Six months later negotiations with Mexican General Vallejo ended with the ceding of property rights and the renaming of the area as San Francisco, after the General's wife, Francisca. When Sam Brannan arrived in the bay on July 31, 1846, bringing with him 230 Mormon immigrants from the east coast, the population more than doubled.

Montgomery's flag raising took place near the adobe Mexican customhouse, at the spot now known as Portsmouth Square, after Montgomery's ship, the U.S.S. Portsmouth. This site, now known as the "Heart of Chinatown," was the center of early community activities and the site of many public meetings and of the first school building. It was there, on May 11, 1848, that Sam Brannan made the dramatic announcement that gold had been discovered.

Portsmouth Square was also the site of the Bella Union Theater, which featured plays, musicals, and cabarets. The square was also where one could find the city's gamblers, prostitutes, and all sorts of bad characters. Along side the Bella Union was the El Dorado, which was a gambling saloon owned by John Piper. He would make his fortune in Virginia City, Nevada, where his Pipers Opera house featured all sorts of entertainment, much like the Bella Union.

In 1882 Fredrick O. Layman built a pavilion high atop Telegraph Hill. Designed as a castle, it contained restaurants, long saloons, and bars. It came to be called "Layman's Folly," because it produced little business and people didn't visit it. To improve business Layman organized a Telegraph Hill Cable Railroad Company and laid tracks from Powell Street to the castle's front door. Finally Adolph Sutro took over the castle, but he too failed to make the resort a success. Over the years it had various incarnations as an athletic club, a vaudeville theater, an artist studio, and was even a bunkhouse for some employees. In 1903 fire broke out and the castle burnt to the ground before a single horse-drawn fire engine could reach it.

One piece of early San Francisco history came to light during the constructing of the Transamerica Building. The whaling ship "Niantic" was built in Connecticut in 1833. Like many "Gold Rush" ships its crew abandoned it when it reached San Francisco in 1849, as they sought their fortunes in the mines. As often happened the abandoned ship was hauled ashore and brought to the corner of Clay and Sansome Streets, which at the time was near the shore. There she was converted into retail space and warehouse, and was used as such until the fire of 1851 burned the ship to her mud line. It was covered over by another building, and later by the filling in of the Yerba Buena Cove. During excavations for the new building in 1977 the hull of the ship was uncovered, and unexpectedly it was found that the lower portions of the hull contained dozens of cases of French wine. Many turned out to be watery mud, but others were still intact. The hull's belly also contained the stock of a stationary store with pencils, paper clamps, and other assorted items from the period. Since this was a construction sight, not an archeological dig, there was very limited time to take and protect these remains, but archeologists saved what they could.

In 1861, as war clouds loomed before the Civil War, Fort Point was built. It was hoped that it would protect the city from an enemy, but the enemy never came.

During the Civil War a young photographer named Carlton Watkins, who was twenty-one, left New York and headed to California, where he landed a job as a Daguerreotype operator in San Jose. He later opened his own studio and photographed around San Francisco and neighboring areas from 1860 to 1862. Then he began traveling around California's Yosemite Valley. Watkins took mammoth plate images of the area and more than one hundred stereo photographs. These were the first photos the eastern United States had ever seen of the area.

Watkins intended to give his entire collection of negatives to Stanford University, but the earthquake and fire destroyed everything. Several Watkins photographs from our collection are included in this book.

Photographer I.W. Tabor was more fortunate. He was able to save some of the negatives and place his own stamp on the views.

On May 10, 1869, the great railroads met at Promontory Point, Utah, starting a national celebration of the rail link across the United States. The Union Pacific and the Central Pacific railroads hammered down the golden spike to mark the occasion.

In San Francisco it was a time for parades. The stores were closed and before long the celebrants hit the saloons and bars. Soon the city was turned into one huge bar with drinks and drunkard's ruling the streets.

Because of the wealth that the mining industry had given bankers and investors, it was a natural choice to build a mint to house all the riches. The San Francisco Mint would be one of the few buildings that held off the great fires in 1906.

For more than thirty years the mining boom kept speculators buying and selling daily, and every day some lost fortunes while others made them. At the center of this financial activity was The Auction House Saloon, run by James C. Flood and William L. O'Brien. From their vantage point Flood and O'Brien would listen for tips on the Comstock share's market and invest accordingly. Their shrewd maneuvering led to their becoming directors of the Bank of Nevada.

When bankers Henry Wells and William G. Fargo opened for business as proprietors of the Pony Express, they thought little of the soon to be completed transcontinental railroad. Their fireproof building on Montgomery Street looked huge, and bore no sign that in a few short years the Pony Express would end its business in favor of stagecoach travel and the railroads. Not to be deterred, they soon opened branches in every mining town in California and Nevada.

By 1870 San Francisco had become the tenth largest city in the United States. Everywhere one looked, new hotels, restaurants, parks, churches, synagogues, schools, libraries, and academies were being built.

It was ironic that Chinese laborers, who were recruited to help build the railroads and work as laborers in the mines and farms, were suddenly turned away. In 1882 Congress passed the Chinese Exclusion Act, barring all Chinese immigration. For the next sixty years those Chinese immigrants who had not registered before then fought their deportation back to China.

There was a time in San Francisco's history when it was known for the making and selling "steam beer" by the Anchor Breweries. This beer was light and had plenty of foam at the top. Served chilled it was certainly a refreshing quaff, but the foam at the top made drinking it a unique experience.

In the 1880s and 1890s the Cliff House at the Seal Rocks was a favorite weekend outing for anyone who had a horse and buggy or a motorcar. It was a place of constant partying and drinking, and had such an enduring appeal that, although it burned down three times, it was rebuilt each time with more luxury than before. Every postcard photographer must have taken some views of this fine structure, even while it was burning down!

The landmark Ferry Building has served transport ships and harbor boats for almost one hundred fifty years, and it too has been rebuilt several times over. People from around the nation were trying get to the gold fields to establish their claims, would rush to the Ferry Building to find passage to Sacramento and beyond. The ferry service began in 1850 with the establishment of a route between San Francisco and Oakland. A group called "The Kangaroo" managed it, and they were so unorganized that the ferryboats seldom met their schedules. Finally in 1852 Oakland granted a license to what was to be the first Bay ferry boat. Travel reached its peak in the 1930s.

An area called West Portal was part of the land holdings of the Mission de Dolores. In 1846, after the break up of the missions, Jose de Noe was granted a 4443-acre ranch, which he called the Rancho San Miguel. In 1880 Adolph Sutro purchased the ranch, which at that time consisted of wide-open, quite hilly spaces. It was named West Portal and, except for one of the flatter portions that was rented by Italian farmers for agricultural purposes, for the next 35 years the rest of San Miguel remained an undeveloped nature preserve. Sutro became the area's tree planter, covering the slops of Mt. Sutro, Mt. Davidson, and as far south as Ocean Avenue.

In 1909, following the earthquake, there was a legal fight over the San Miguel. While the Burnham plan of 1905 and the City Beautiful movement called for respecting the contours of the land and to incorporate landscaping into new residential developments, development was inevitable.

The area bounded by Montgomery, Stockton, Washington and Broadway, was called the Barbary Coast. In daylight hours one would see simple, ordinary, day-to-day people doing their everyday jobs, but at night it would become another world. During the day, Chinese porters could be seen swinging heavy burdens on long bamboo poles. You would see Chinese females, clad in loose fitting clothing, and shoes of white felt and wood. Their hair would be braided in two strands thrown over their heads and tied beneath their chin as a badge denoting slavery. Every second building was operated as a saloon, but inside no one stirred during the day. The buildings were dilapidated and dirty, and from all appearances, nothing was going on.

But at night, and specially during the rainy season when no work is being done at the ranches or at the mines, the area swarmed with miners, unemployed farm hands, jay-hawkers, short card sharps, pick pockets, prostitutes, and their victims.

In 1871 this part of San Francisco was a world of it's own. And it drew people of high fashion and wealth who would parade around the streets, looking into places they would not dare enter. The club rooms beckoned them to enter, the saloons all looked neat and orderly, and even the fellow at the rear door of a club room in the back was very polite, but he stopped anyone he didn't know from entering.

On Pacific Street, alone, there were 48 cabarets, saloons, and resorts in 1905. They bore such names as, Zaza Café, Oregon Dance Hall, House of all Nations, So Different Club (for African Americans), Wine Joint, and The Dixie, to name a few.

Add horse and auto races to the clubs and gambling houses, and it is easy to see that San Francisco was a lively city. One could easily become lost in the sea of entertainment opportunities. But despite all of this, or perhaps because of it, the city government was totally corrupt, and its leaders were filling their pockets with the gold of bribes and kickbacks.

Mayor Eugene Schmitz and his public accountant, Abe Ruef, kept tabs on all the city spending. This power of the purse strings, gave him great control, including the hiring of police and fire personnel, the purchasing of fire equipment, and even the adding of water tanks to the tops of the larger buildings. He would not bend to groups who wanted their city to be a safe place to live.

This hegemony was challenged in the courtroom of Judge W.W. Morrow, a short time before the earthquake. The Judge, Mayor Schmitz, and several civic leaders met to discuss a 1905 report of the National Board of Fire Underwriters, and to decide on steps to make the city safer. The report was critical of the status of the city, stating "San Francisco has violated all underwriter's traditions by not burning up." That this hadn't happened was largely due to the work of the city's fire department.

But Chief Sullivan knew that his efforts were not enough to ward off a major fire, and he had a long list of civic improvements. He was in dire need of money to obtain the proper equipment needed to fight large fires in the city. He also wanted money for a supplementary salt water system for hydrants and to activate scores of neglected cisterns that were laid beneath the streets a dozen years before. Finally, he wanted money to train his men on the use of high explosives to check major fires. He made it clear that the Underwriters could only report the problem; they couldn't fix it.

The result of the meeting? Nothing. After listening to the Underwriters report, the urging of his fellow citizens, and Chief Sullivan, City hall refused to act on the any of the recommendations. Mayor Schmitz, it seems, liked the way things had been managed by his associate Abe Ruef. But the end was in sight. After all these years of grafting money from everyone in the city, Schmitz's days as mayor were numbered. He learned that millionaire Rudolph Spreckles had launched an investigation of city hall and even paid for a Washington, D.C. Secret Service agent William J. Burns, to gather evidence of bribery and corruption by Ruef and Schmitz.

The Fairs were one of the leading families of San Francisco before the 1906 earthquake. James Graham Fair had struck it rich with a Nevada silver mine. After his death, his daughters, Tessie and Virginia, planned to build a grand hotel in San Francisco as a monument to him. Construction began in 1902. By 1906 it was finished. Everything was in place, including the furniture and paintings. But the project had become too much for the sisters, so they sold the nearly complete hotel to the brother Herbert and Hartland Law.

The sisters' timing was prescient. Within a few days of the sale, the earthquake hit and the hotel was severely damaged by fire...before hosting a single guest. Not to be deterred, the Law brothers mustered their energies and restored the grand hotel. It became a center piece of the San Francisco scene, hosting every U.S. President since William Howard Taft.

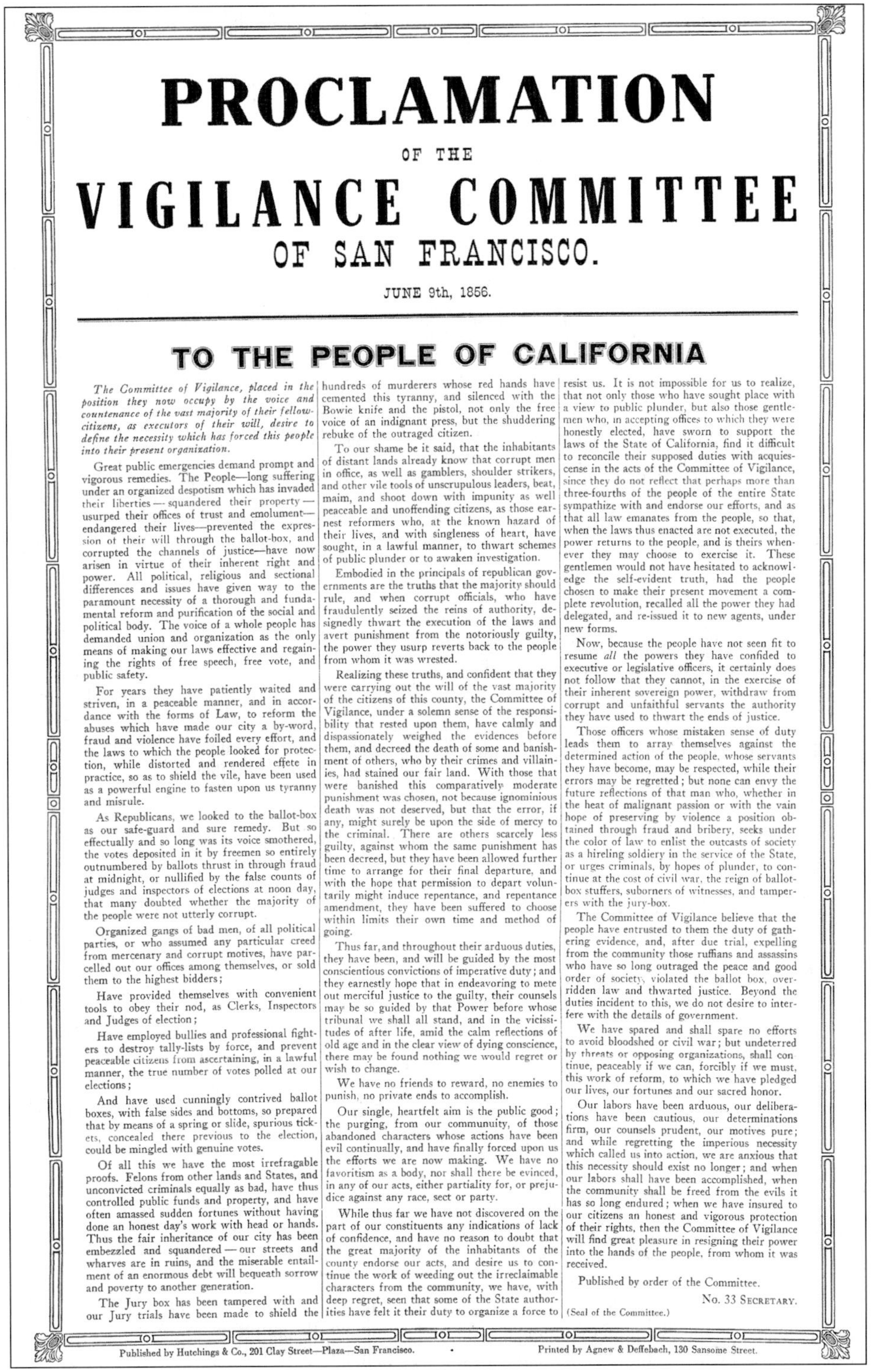

# PROCLAMATION

OF THE

# VIGILANCE COMMITTEE

OF SAN FRANCISCO.

JUNE 9th, 1856.

## TO THE PEOPLE OF CALIFORNIA

*The Committee of Vigilance, placed in the position they now occupy by the voice and countenance of the vast majority of their fellow-citizens, as executors of their will, desire to define the necessity which has forced this people into their present organization.*

Great public emergencies demand prompt and vigorous remedies. The People—long suffering under an organized despotism which has invaded their liberties — squandered their property — usurped their offices of trust and emolument—endangered their lives—prevented the expression of their will through the ballot-box, and corrupted the channels of justice—have now arisen in virtue of their inherent right and power. All political, religious and sectional differences and issues have given way to the paramount necessity of a thorough and fundamental reform and purification of the social and political body. The voice of a whole people has demanded union and organization as the only means of making our laws effective and regaining the rights of free speech, free vote, and public safety.

For years they have patiently waited and striven, in a peaceable manner, and in accordance with the forms of Law, to reform the abuses which have made our city a by-word, fraud and violence have foiled every effort, and the laws to which the people looked for protection, while distorted and rendered effete in practice, so as to shield the vile, have been used as a powerful engine to fasten upon us tyranny and misrule.

As Republicans, we looked to the ballot-box as our safe-guard and sure remedy. But so effectually and so long was its voice smothered, the votes deposited in it by freemen so entirely outnumbered by ballots thrust in through fraud at midnight, or nullified by the false counts of judges and inspectors of elections at noon day, that many doubted whether the majority of the people were not utterly corrupt.

Organized gangs of bad men, of all political parties, or who assumed any particular creed from mercenary and corrupt motives, have parcelled out our offices among themselves, or sold them to the highest bidders;

Have provided themselves with convenient tools to obey their nod, as Clerks, Inspectors and Judges of election;

Have employed bullies and professional fighters to destroy tally-lists by force, and prevent peaceable citizens from ascertaining, in a lawful manner, the true number of votes polled at our elections;

And have used cunningly contrived ballot boxes, with false sides and bottoms, so prepared that by means of a spring or slide, spurious tickets, concealed there previous to the election, could be mingled with genuine votes.

Of all this we have the most irrefragable proofs. Felons from other lands and States, and unconvicted criminals equally as bad, have thus controlled public funds and property, and have often amassed sudden fortunes without having done an honest day's work with head or hands. Thus the fair inheritance of our city has been embezzled and squandered — our streets and wharves are in ruins, and the miserable entailment of an enormous debt will bequeath sorrow and poverty to another generation.

The Jury box has been tampered with and our Jury trials have been made to shield the hundreds of murderers whose red hands have cemented this tyranny, and silenced with the Bowie knife and the pistol, not only the free voice of an indignant press, but the shuddering rebuke of the outraged citizen.

To our shame be it said, that the inhabitants of distant lands already know that corrupt men in office, as well as gamblers, shoulder strikers, and other vile tools of unscrupulous leaders, beat, maim, and shoot down with impunity as well peaceable and unoffending citizens, as those earnest reformers who, at the known hazard of their lives, and with singleness of heart, have sought, in a lawful manner, to thwart schemes of public plunder or to awaken investigation.

Embodied in the principals of republican governments are the truths that the majority should rule, and when corrupt officials, who have fraudulently seized the reins of authority, designedly thwart the execution of the laws and avert punishment from the notoriously guilty, the power they usurp reverts back to the people from whom it was wrested.

Realizing these truths, and confident that they were carrying out the will of the vast majority of the citizens of this county, the Committee of Vigilance, under a solemn sense of the responsibility that rested upon them, have calmly and dispassionately weighed the evidences before them, and decreed the death of some and banishment of others, who by their crimes and villainies, had stained our fair land. With those that were banished this comparatively moderate punishment was chosen, not because ignominious death was not deserved, but that the error, if any, might surely be upon the side of mercy to the criminal. There are others scarcely less guilty, against whom the same punishment has been decreed, but they have been allowed further time to arrange for their final departure, and with the hope that permission to depart voluntarily might induce repentance, and repentance amendment, they have been suffered to choose within limits their own time and method of going.

Thus far, and throughout their arduous duties, they have been, and will be guided by the most conscientious convictions of imperative duty; and they earnestly hope that in endeavoring to mete out merciful justice to the guilty, their counsels may be so guided by that Power before whose tribunal we shall all stand, and in the vicissitudes of after life, amid the calm reflections of old age and in the clear view of dying conscience, there may be found nothing we would regret or wish to change.

We have no friends to reward, no enemies to punish, no private ends to accomplish.

Our single, heartfelt aim is the public good; the purging, from our communuity, of those abandoned characters whose actions have been evil continually, and have finally forced upon us the efforts we are now making. We have no favoritism as a body, nor shall there be evinced, in any of our acts, either partiality for, or prejudice against any race, sect or party.

While thus far we have not discovered on the part of our constituents any indications of lack of confidence, and have no reason to doubt that the great majority of the inhabitants of the county endorse our acts, and desire us to continue the work of weeding out the irreclaimable characters from the community, we have, with deep regret, seen that some of the State authorities have felt it their duty to organize a force to resist us. It is not impossible for us to realize, that not only those who have sought place with a view to public plunder, but also those gentlemen who, in accepting offices to which they were honestly elected, have sworn to support the laws of the State of California, find it difficult to reconcile their supposed duties with acquiescense in the acts of the Committee of Vigilance, since they do not reflect that perhaps more than three-fourths of the people of the entire State sympathize with and endorse our efforts, and as that all law emanates from the people, so that, when the laws thus enacted are not executed, the power returns to the people, and is theirs whenever they may choose to exercise it. These gentlemen would not have hesitated to acknowledge the self-evident truth, had the people chosen to make their present movement a complete revolution, recalled all the power they had delegated, and re-issued it to new agents, under new forms.

Now, because the people have not seen fit to resume *all* the powers they have confided to executive or legislative officers, it certainly does not follow that they cannot, in the exercise of their inherent sovereign power, withdraw from corrupt and unfaithful servants the authority they have used to thwart the ends of justice.

Those officers whose mistaken sense of duty leads them to array themselves against the determined action of the people, whose servants they have become, may be respected, while their errors may be regretted; but none can envy the future reflections of that man who, whether in the heat of malignant passion or with the vain hope of preserving by violence a position obtained through fraud and bribery, seeks under the color of law to enlist the outcasts of society as a hireling soldiery in the service of the State, or urges criminals, by hopes of plunder, to continue at the cost of civil war, the reign of ballot-box stuffers, suborners of witnesses, and tamperers with the jury-box.

The Committee of Vigilance believe that the people have entrusted to them the duty of gathering evidence, and, after due trial, expelling from the community those ruffians and assassins who have so long outraged the peace and good order of society, violated the ballot box, overridden law and thwarted justice. Beyond the duties incident to this, we do not desire to interfere with the details of government.

We have spared and shall spare no efforts to avoid bloodshed or civil war; but undeterred by threats or opposing organizations, shall continue, peaceably if we can, forcibly if we must, this work of reform, to which we have pledged our lives, our fortunes and our sacred honor.

Our labors have been arduous, our deliberations have been cautious, our determinations firm, our counsels prudent, our motives pure; and while regretting the imperious necessity which called us into action, we are anxious that this necessity should exist no longer; and when our labors shall have been accomplished, when the community shall be freed from the evils it has so long endured; when we have insured to our citizens an honest and vigorous protection of their rights, then the Committee of Vigilance will find great pleasure in resigning their power into the hands of the people, from whom it was received.

Published by order of the Committee.

No. 33 SECRETARY.

(Seal of the Committee.)

Published by Hutchings & Co., 201 Clay Street—Plaza—San Francisco. • Printed by Agnew & Deffebach, 130 Sansome Street.

A June 9, 1856 proclamation of the Vigilance Committee of San Francisco to the people of California, stating that "organized gangs of bad men...have parceled out our offices among themselves and sold them to the highest bidders...Felons from other lands and states, and unconvicted criminals just as bad, have...controlled public funding...Thus the fair inheritance of our city has been embezzled and squandered—our streets and wharves are in ruins." It was all out war in San Francisco.

The following six photos show San Francisco in 1862. They were printed by L. Nagel, lithographed by C. Gifford, and published by A. Rosenfield, San Francisco. The views are looking west from Russian Hill.

Telegraph Hill is seen center left. The island rear center is Goat Island.

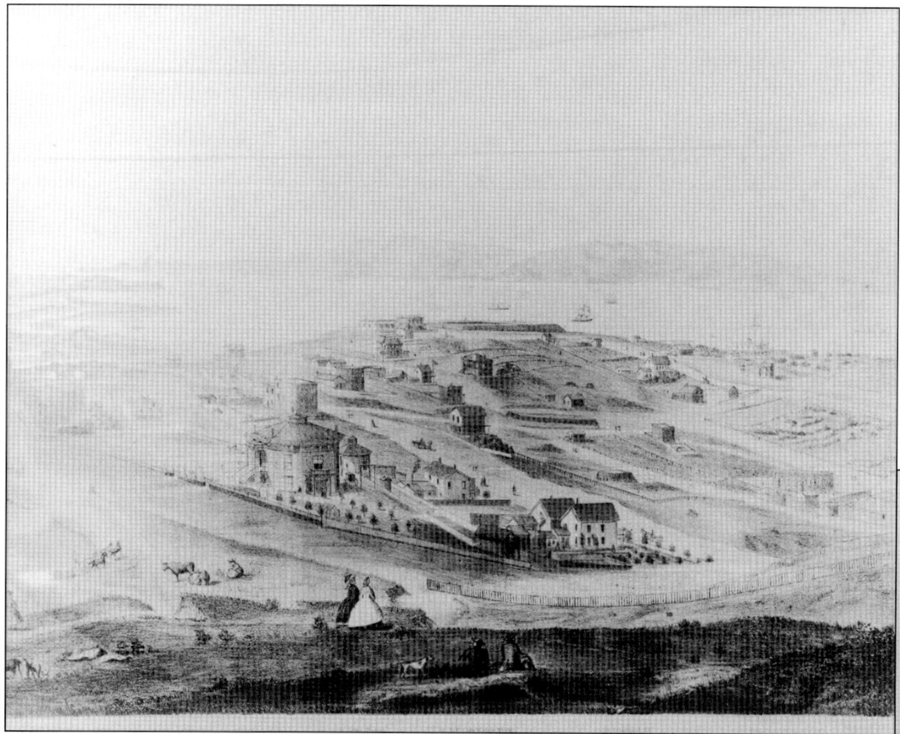

At the left of center is the Water Works building.

Alcatraz Island is seen in the center and Meiggs Wharf is to the right.

An over all view of a thriving city, showing homes, commercial buildings, churches, and more.

The Clay Street tower the houses surrounding it.

A view of harbor area with homes, wharfs and other buildings.

An early San Francisco ferry boat makes its way to the dock at the Vallejo and Davis street landing. The photograph, taken around 1878, also shows several three-masted schooners and another ferry boat tied up.

A common sight in early San Francisco was the horse drawn streetcar, captured here on an early stereo card photograph. The scene shows Montgomery street from Market street

No. 5 Market Street east of Fifth Street, showing an active, busy city.

A lovely midwinter's day at Golden Gate park, where people would come out for a day's rest and perhaps a picnic.

A Clay Street Railroad car is seen at Leavenworth Street Hill in 1878.

Mt. Tamalpais and Alcatraz Island as seen from Telegraph Hill in an original Carlton Watkins photograph.

Golden Gate from Telegraph Hill. An original photograph by Carlton Watkins.

California Street, from Sansome, from an original Carlton Watkins photograph.

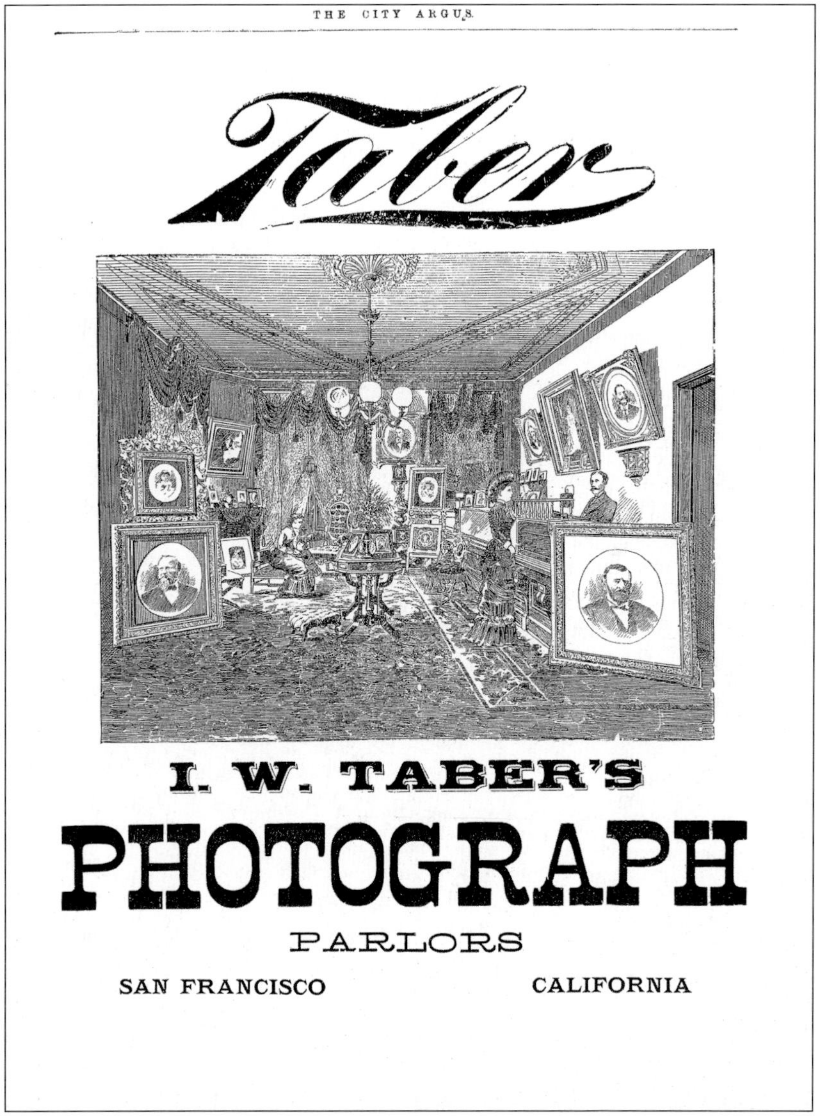


A *City Argus* advertisement for I. W. Tabor's Photo studio, in 1893.

Photographer I. W. Tabor took his camera to the horse races and recorded this fine example of an early "action" photograph. Since film at this time was quite slow and needed strong light, this image is quite dark, but the remarkable thing is that Tabor captured it at all.

The original San Francisco City Hall. It was destroyed by fire in 1876.

An early auto race car zooms into the camera's view , and in the background sits a 1903 automobile. This action photograph is by an unknown photographer.

A typical Fire Station in early San Francisco showing the "Rescue Steamer Number 2." This photo is from an original albumin print.

With a Sunday off, "Rescue Steamer Number 3's" crew stands for their portrait.

M. MEUSSDORFFER'S
Spring and Summer Styles
Of Dress and Business

HATS

Are the ones which Every Gentleman of Taste Should Wear.

They are appreciated by calling at his Modern Store,

113 MONTGOMERY STREET
Opp. Occidental Hotel

Wonder! Wonder!
LADIES, CALL AT THE
WONDER
Hat, Flower and Feather Store,
1024, 1026 & 1028 Market St.
AND SEE OUR LARGE STOCK OF
SPRING NOVELTIES!
—IN—
Hats, Flowers, Laces, Ribbons
ETC., ETC. ETC.
LARGE STOCK! LOW PRICES! NEW STYLES!
WILL NOT BE UNDERSOLD!

KEELEY INSTITUTE
Los Gatos, Cal.
O. N. RAMSEY, MANAGER.

The only authorized branch of the world renowned Keeley Institute of Dwight, Ill., in California.
Certain cure for Liquor, Opium, Tobacco and Nervous diseases.
Parties in San Francisco and Vicinity can obtain all nformation by calling on or addressing
PEER TIFFANY, City Agent,
Room 37, Academy of Sciences Building
819 MARKET STREET.

RICHARD T. CONDON
—WITH—
Manheim Dibbern & Co.
FIRE INSURANCE AGENTS
217 SANSOME STREET.

MATTHIAS GRAY CO.
206 POST STREET, S. F.
General Agents for
STEINWAY & SONS' PIANOS
ERNEST GABLER PIANOS
KRANICH & BACH PIANOS
C. ROENISCH PIANOS
PEASE PIANOS
PACKARD ORGANS
Importers of American and Foreign
Band Instruments, Accordeons, Violins, Guitars, Sheet Music, Books, Etc.
Everything in the Music Line

W. G. BADGER
Sole Agent for
HALLET, DAVIS & CO.
FRANCIS BACON
W. W. KIMBALL
AND GILMAN
PIANOS
—AND—
W. W. KIMBALL
ORGANS
721 Market St. San Francisco

BOLTON & STRONG
MANUFACTURERS OF
Printing Plates
Half-tones
Line Engravings
Etc.
430 Pine Street San Francisco
Estimates furnished on application.

THE BURLINGTON
605 Market St. Cor. Second
SAN FRANCISCO
New and Elegantly Furnished Rooms en Suite or Single at Reasonable Prices.
Traveling Public Solicited.
MRS. BURLING,
Elevator. PROPRIETRESS

GERMEA
The Monarch of Breakfast Foods
THE JOHN T. CUTTING CO, SOLE AGENTS

Wells, Fargo & Co's
—BANK—
N.E. cor. Sansome & Sutter Sts.
SAN FRANCISCO.
Cash Capital and Surplus .. $6,250,000.00
JOHN J. VALENTINE..........President
HOMER S. KING..........Manager
H. WADSWORTH..........Cashier
F. L. LIPMAN..........Assistant Cashier
DIRECTORS:
John J. Valentine, Lloyd Tevis, Leland Stanford, Oliver Eldridge, James C. Fargo, Geo. E. Gray, Chas. F. Crocker, W. F. Goad, Dudley Evans.
Receive Deposits, issues Letters of Credit, and transacts a General Banking Business.

Vichy Springs
THREE MILES FROM UKIAH
Only abundant supply of Vichy Water in the United States. The well-known "Champagne" Baths. The qualities of softening and beautifying the skin testified to by hundreds of ladies. Only natural electric water in the world.
☞ Write for analysis and consult your physician.
WILLIAM DOOLAN,
PROPRIETOR

ESTABLISHED 1850
Kohler & Chase
PIANO & MUSIC WAREROOMS
137 and 139 POST ST., San Francisco
Dealers in Band and all other
Musical Instruments
Low Prices Cash or Easy Installments
Tuning, Polishing and Repairing Done. First-Class Workmen and Satisfaction Guaranteed.
BRANCH STORES:
OAKLAND, CAL. PORTLAND, OR

JOHN WIELAND
FREDERICKSBURG
UNITED STATES
CHICAGO
Breweries
Lager Beer Bottled Expressly for Family Use.
Gen'l Office, 403 Market Street
TELEPHONE 1150
SAN FRANCISCO CALIFORNIA

Bartlett
Spring
Mineral - Water
ACKNOWLEDGED BY PHYSICIANS to be the Greatest Mineral Water for the Cure of Constipation, Kidney and Liver Troubles, and unsurpassed for the Cure of Dyspepsia, Rheumatism, Gout, Dropsy, St. Vitus Dance, Chronic Alcoholism and Blood Diseases.
BARTLETT SPRINGS COMPANY
22 Fourth Street (Basement)
PIONEER BUILDING SAN FRANCISCO

[ESTABLISHED 1879]
OTTO NORMANN
411 BUSH ST., Opp. the New California Hotel
BEST EQUIPPED
Oyster & Cold Lunch Parlors
Large Dining Room for Ladies
SOLE DEPOT FOR
JOS. SCHLITZ MILWAUKEE BEER

Sanborn, Vail & Co.
857—MARKET ST.—857
MANUFACTURERS OF
MOULDINGS
FRAMES,
MIRRORS
—AND—
ARTISTS' MATERIALS

DR. WM. F. EGAN,
M. R. C. V. S., F. E. V. M. S
VETERINARY SURGEON
Member of the Royal College of Veterinary Surgeons, England; Fellow of the Edinburgh Veterinary Medical Society; Graduate of the New Veterinary College, Edinburgh; Ex-Veterinary Surgeon to the Board of Health for the City and County of San Francisco; Member of the California State Veterinary Medical Association; Veterinary Editor to *Breeder and Sportsman*.
Veterinary Infirmary, Residence and Office:
1117 GOLDEN GATE AVE.,
Telephone No. 4128, - - SAN FRANCISCO
BRANCH OFFICES
1525 California Street, Telephone 66.
1529 Howard Street, Telephone 3153.

An advertising page in *City Argus* newspaper in 1893

A very rare photograph of the interior of a black club with dancers and a band at the Barbary Coast. Nellie Lewis who signed this photo appeared in black minstrel shows across the country. Notice the strong oriental designs on the walls. There were only three Black Clubs listed in the city directory of the period.

Nellie Lewis
"Barbary Coast"

A horse-drawn streetcar stops for its portrait along with people who lived nearby in Ross Garden. This image is from a stereo photograph.

Photographer J. McDonald took this wonderful view of "Life in Golden Gate Park."

An early view of San Francisco Harbor and Alcatraz Island in the background.

A very rare photograph of the Telegraph Hills Observatory, with a streetcar waiting for passengers.

This street scene shows McAllister Street at Fillmore, looking west, in San Francisco.

San Francisco Harbor with several four-masted ships. The image appeared on an early post-card.

The second San Francisco City Hall, as it looked before the 1906 earthquake and fires.

The left side of this wonderful pen and ink drawing by Carl Browne, Esq. Pub. City Argus.

The right side of the "Tour Around The World" art By Carl Browne, Esq. City Argus.

A Chinese fortune teller and letter writer, at his stand.

A 1903 panoramic view of San Francisco. The Claus Spreckles Building rises above the others in the center of the scene. It was the tallest building on the west coast and home to *The Call* newspaper. Rising 16 stories above the street, its dome had a restaurant with a tremendous perspective of the city.

This bird's-eye view was from the Fairmont hotel before the earthquake. It is taken toward Oakland, and shows the Ferry Building in the background.

Photographer Frank Warner of San Francisco took this action shot of a 2-year-old horse trotting.

With not much to do on a Sunday afternoon, crowds of people took to walking along the shores of the beach near the Cliff House.

A view taken from another angle showing the Cliff House in the distance and the crowds on the beach watching Seal Rock.

A stage taking a group from San Francisco to some of the mines in the hills.

From a *Harpers* magazine of the period, this drawing shows police searching Chinese immigrants for opium.

The San Francisco police line up for a full dress inspection at Golden Gate Park. Though unidentified, one can only assume that the gentleman in the top hat was an important dignitary.

A policeman holds still for his portrait by an unknown photographer.

San Francisco was an important port for shipping, as this 1903 photograph shows. A typical harbor scene, one would find the docks overflowing with materials coming and going.

In 1897 the Sutro Bath House was a big hit, with hundreds of bathers lining the edges of the pools.

Another interior view of Sutro's Baths. The pools were designed so that each pool had a different temperature.

Sutro himself poses below his bathhouse, as guests look down at him.

Fort Point at the Golden Gate from a early postcard.

The famous U.S. Mint in San Francisco that withstood the earthquake and great fire.

The seals on Seal Rock near the Cliff House.

Chinatown's Fish Alley looked like this before the earthquake. The fire that followed destroyed this entire area.

A scene in Chinatown, before the earthquake.

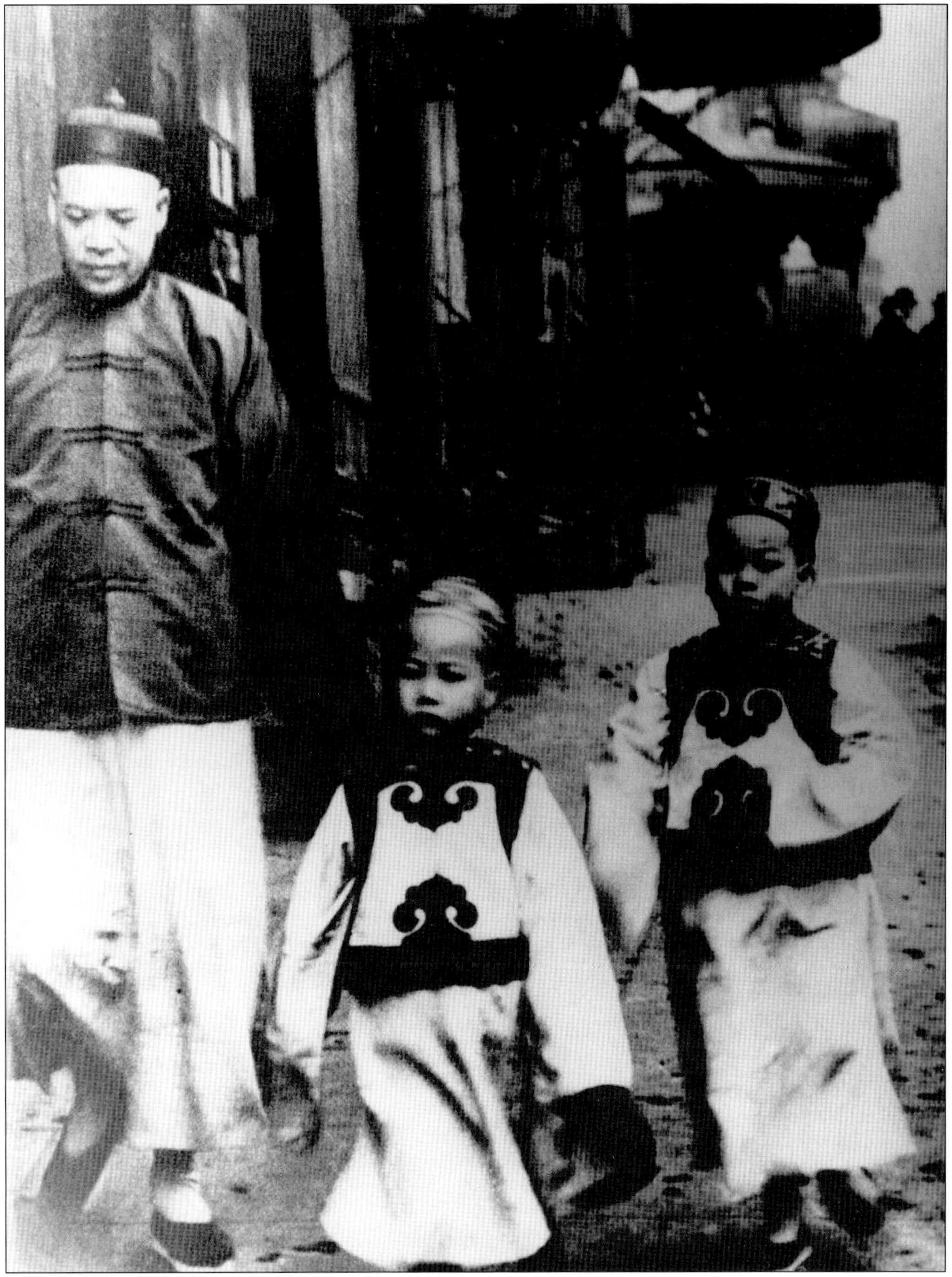

An officer of Chinatown, walks his grand children who are dressed in their traditional costume.

A busy street scene in Chinatown, before the earthquake.

San Francisco's Chinatown was known to have underground opium dens like the one pictured here.

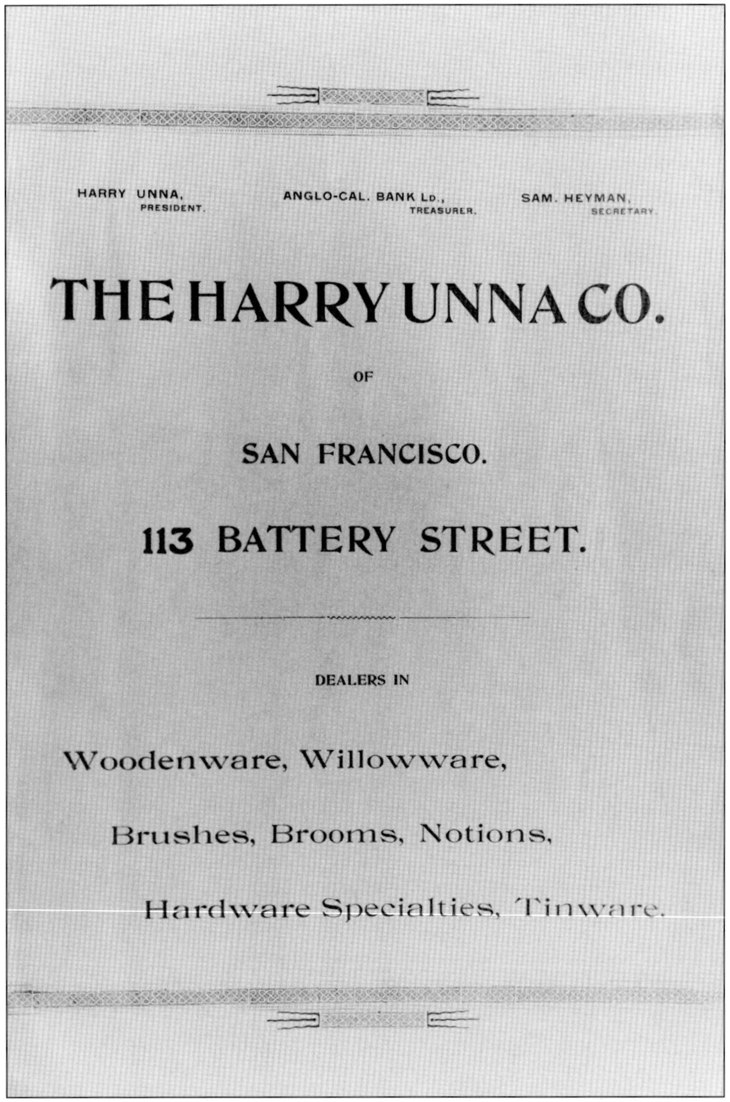
HARRY UNNA, PRESIDENT. ANGLO-CAL. BANK Ld., TREASURER. SAM. HEYMAN, SECRETARY.

THE HARRY UNNA CO.

OF

SAN FRANCISCO.

113 BATTERY STREET.

DEALERS IN

Woodenware, Willowware,

Brushes, Brooms, Notions,

Hardware Specialties, Tinware.

Like other merchants in San Francisco, the Harry Unna Company on Battery Street offered all sorts of things for house and home.

113 BATTERY STREET. 7

THE BEST

Syphon and Pump Oil Can

MADE IN AMERICA.

THE IMPROVED

"NEW MIKADO"

SIMPLICITY OF CONSTRUCTION.
NEATNESS OF WORKMANSHIP.
STRENGTH AND GENERAL UTILITY.
PAR-EXCELLENCE OF APPEARANCE.
REMOVABLE PUMP.
DEPRESSED SAFE TO RECEIVE ALL DRIP.
MILLED SCREW CAP TO KEEP OUT DUST AND PROTECT PARTS.
A PERFECT INNER COATING OR OIL RE-PELLANT.

MADE OF GALVANIZED IRON.

HOLDS 5 GALLONS.

Price.................................$15.00 Per Dozen.



Reciprocity

FAUCET OIL CAN.

(5 Gallons).

*Heavy galvanized iron. nickel plated faucet.*

FOR FAMILY USE.

Oil can be drawn from these cans without any leakage, waste or danger.

Per Dozen............$9.00



A page in the Unna Company's catalog offered a fancy siphon in their large booklet.

Another page in Harry Unna's catalog had shoe brushes boot blacking materials.



## SHOE BRUSHES.

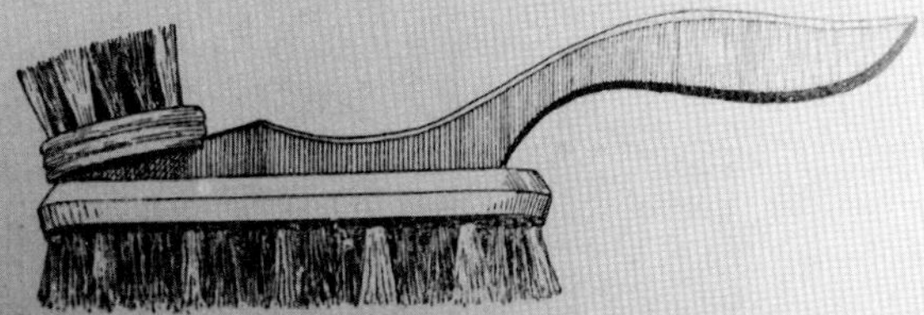

(No. 3.)

| NO. | | PER DOZEN |
|---|---|---|
| 3, | Best Value in the Market. Good Stiff, Black and White Stock | $1.20 |
| 9, | Solid Back, all Black Stock | 1.75 |
| 69, | Fine Selected Black Bristle, Extra Finished Brush, Usually Retailed for 40c | 2.10 |
| 140, | Good Stiff Stock, Very Full | 3.00 |

We have Shoe Brushes from 95c to $6.00 Per Dozen.

## JACQUOT'S FRENCH BLACKING.

| | |
|---|---|
| No 3 | $4.50 Per Gross |
| No. 4 | 5.50 |

(In ¼ Gross Boxes.)

## ROYAL LIQUID BLUING.

PUT UP IN PINT BOTTLES.

Per Dozen ........ 75 Cents

(Two Dozen in a Case.)

We guarantee this Bluing to give satisfaction. Retails for 10 Cents, and gives you a good

A young Chinese man named Sing, poses by his open book.

A young beauty sat for this portrait at Carlton Watkins studio at 22 &26 Montgomery street.

What the well-dressed young man in San Francisco would wear in the 1890s.

This young scholar was named Lock Jim. He sat for this portrait study by a unknown photographer.

This full length portrait of a finely dressed woman was taken at Marden's Art Gallery, No. 28 Third Street, San Francisco.

The Immigrant Station at Angel Island, San Francisco harbor.

SANBORN

This busy street scene captures something of the vibrant, bustling city of San Francisco before the earthquake hit. Streetcars and horse-drawn wagons rushing to and fro through the streets, vying for space with the pedestrians.

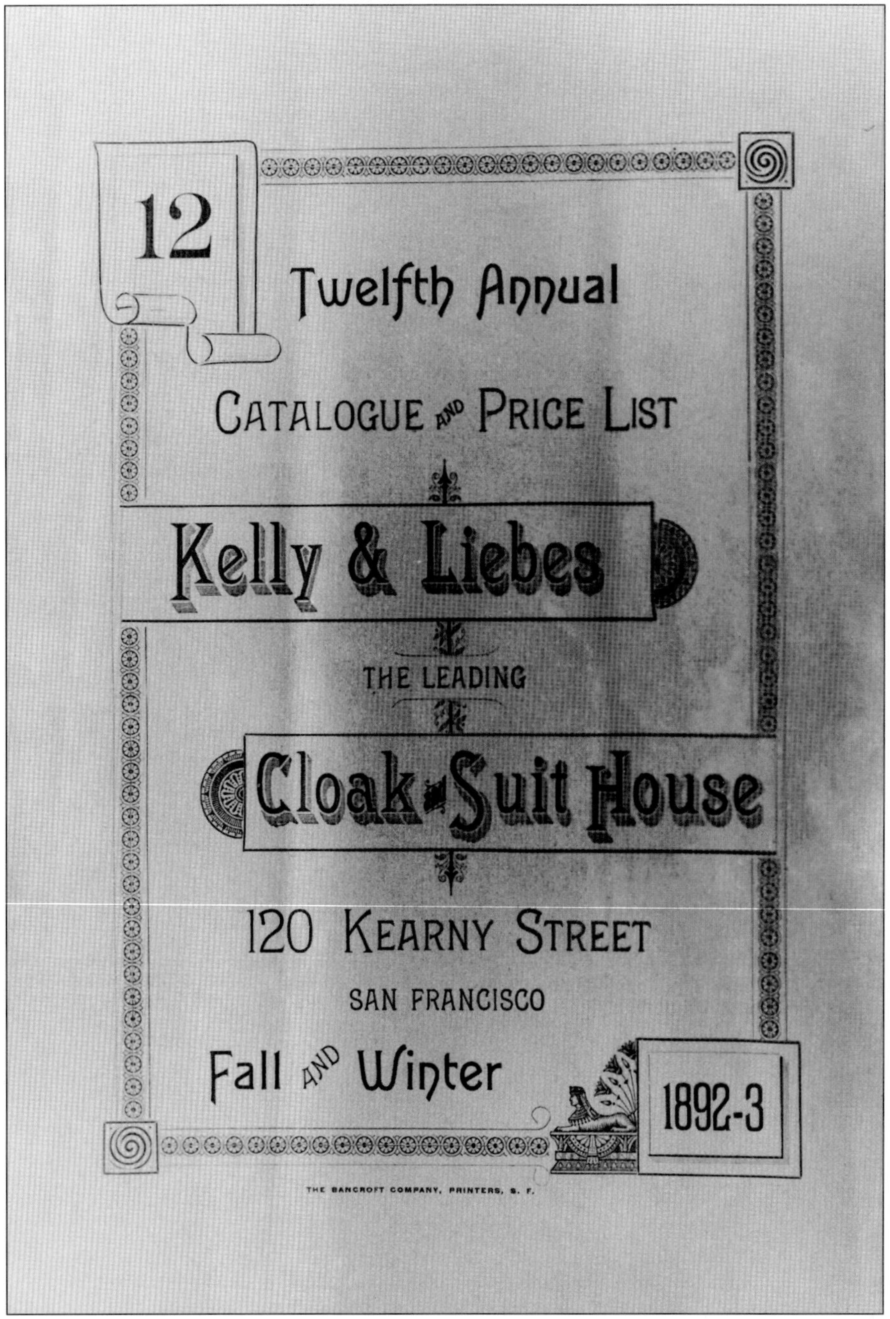


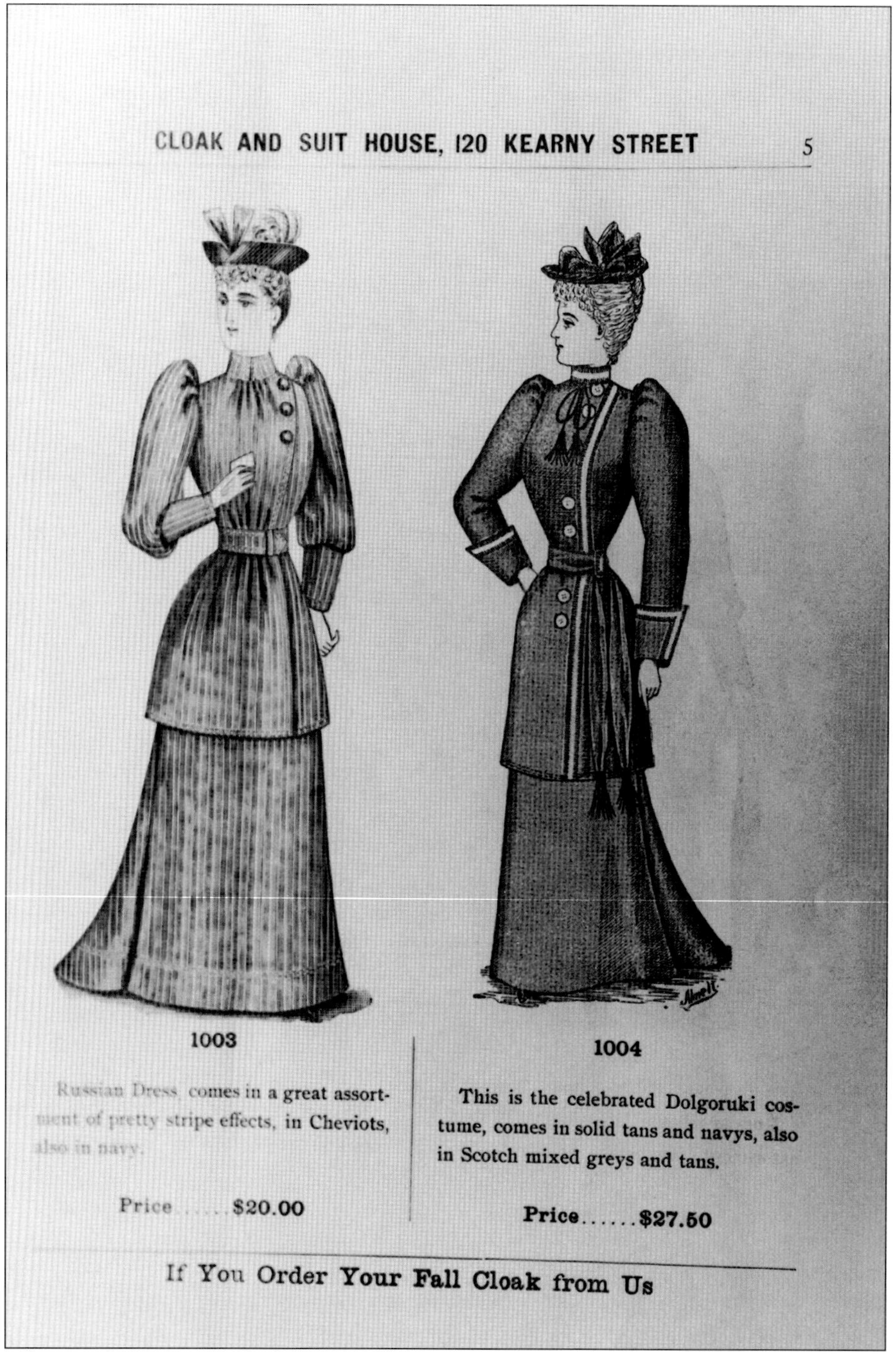


One of the retailers in women's clothes in the 1890s was Kelly & Liebes on Kearney Street, "the leading Cloak and Suit House." Appearing in the 1884 directory, they do not reappear after the earthquake.

Interest in the latest fashions is clear indication of the prosperity of San Francisco's citizens.

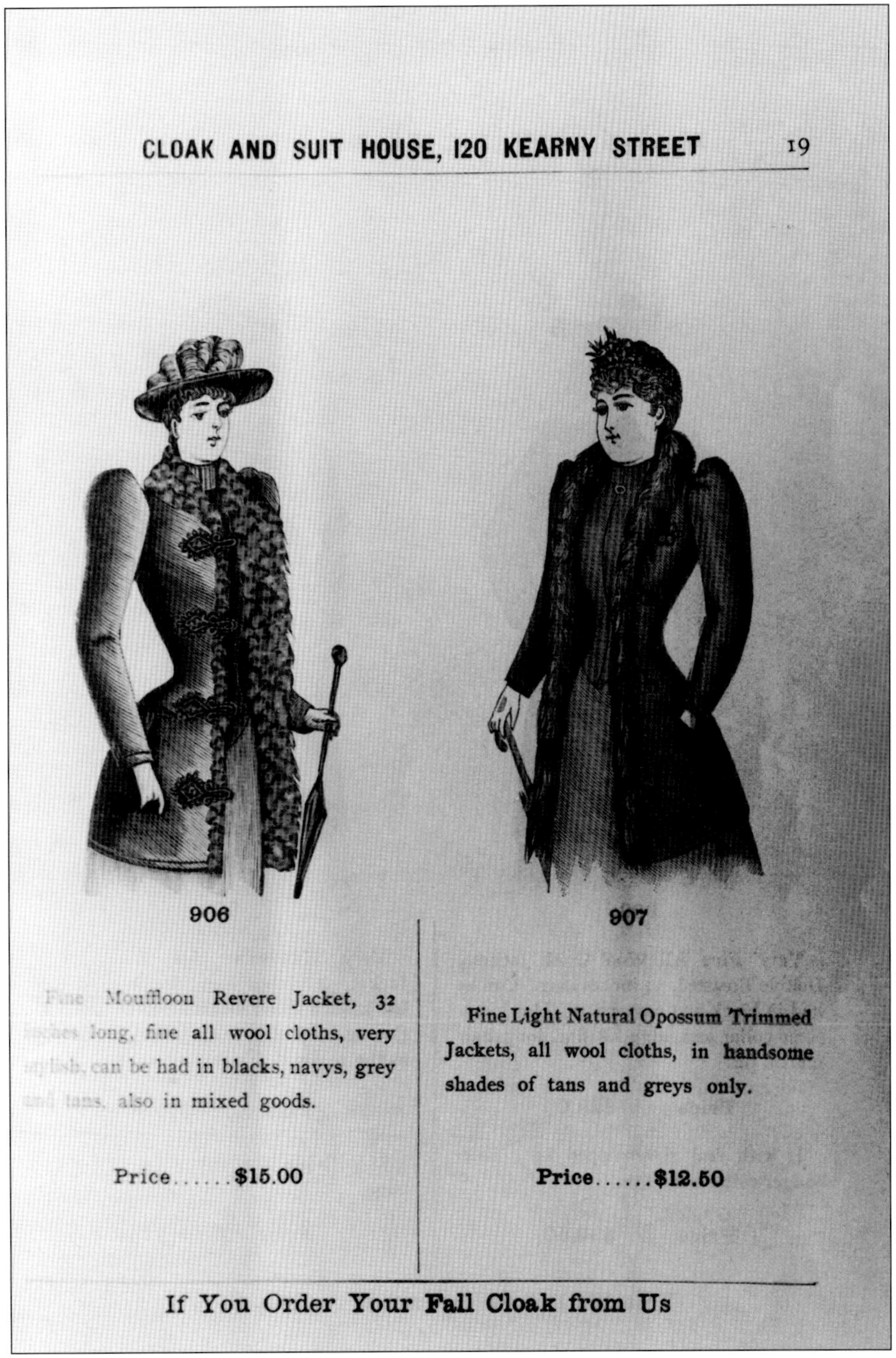

CLOAK AND SUIT HOUSE, 120 KEARNY STREET 19

906

Fine Moufflon Revere Jacket, 32 inches long, fine all wool cloths, very stylish, can be had in blacks, navys, grey and tans, also in mixed goods.

Price......$15.00

907

Fine Light Natural Opossum Trimmed Jackets, all wool cloths, in handsome shades of tans and greys only.

Price......$12.50

If You Order Your Fall Cloak from Us

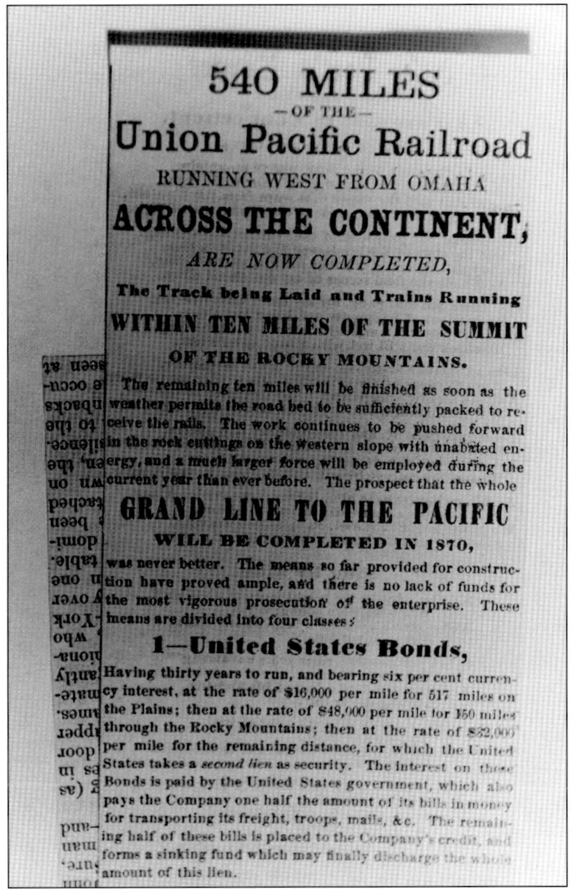

540 MILES

—OF THE—

Union Pacific Railroad

RUNNING WEST FROM OMAHA

ACROSS THE CONTINENT,

*ARE NOW COMPLETED,*

The Track being Laid and Trains Running

WITHIN TEN MILES OF THE SUMMIT

OF THE ROCKY MOUNTAINS.

The remaining ten miles will be finished as soon as the weather permits the road bed to be sufficiently packed to receive the rails. The work continues to be pushed forward in the rock cuttings on the western slope with unabated energy, and a much larger force will be employed during the current year than ever before. The prospect that the whole

GRAND LINE TO THE PACIFIC

WILL BE COMPLETED IN 1870,

was never better. The means so far provided for construction have proved ample, and there is no lack of funds for the most vigorous prosecution of the enterprise. These means are divided into four classes:

1—United States Bonds,

Having thirty years to run, and bearing six per cent currency interest, at the rate of $16,000 per mile for 517 miles on the Plains; then at the rate of $48,000 per mile for 150 miles through the Rocky Mountains; then at the rate of $32,000 per mile for the remaining distance, for which the United States takes a *second lien* as security. The interest on these Bonds is paid by the United States government, which also pays the Company one half the amount of its bills in money for transporting its freight, troops, mails, &c. The remaining half of these bills is placed to the Company's credit, and forms a sinking fund which may finally discharge the whole amount of this lien.

An announcement from the Union Pacific Railroad that 540 miles have been completed for the transcontinental railroad, and that the whole line will be finished in 1870.

A photo of the old Cliff House before several fires burned it to the ground.

The steamer Tamalpas and others at the Sansilto landing.

Hodson's photograph of the U.S. Armies Third Brigade, Eight Army Corps under Brig. Gen. H.G. Otis, at Golden Gate Park. These troops fought in Montana, South Dakota, Iowa, and as the 7th California Regiment.

This portrait of an actress was taken in 1898 by Theo Marceau of 826 Market Street. Marceau had studios in Cincinnati, Ohio, as well as in Indianapolis, Indiana.

The Fourth of July parade in San Francisco always had some dignified-looking people who came along for the ride. Even the policeman had some medals to prove his worth.

A photograph of Hayes street looking west from Octavia Street. Streetcar tracks are being laid down the middle of the street.

# Giant Waves & Falling Buildings

# The Earthquake and Fire

Out in the Pacific Ocean the schooner John A. Campbell shook, her bow rose in the air then crashed down as if hit by a giant wave. The crew leaped to their feet and ran out on deck thinking they had hit another ship or a whale, but nothing was there and the water was still.

The motion they experienced was the result of stress miles below the ocean's surface. Unknown to the crew of the John A. Campbell, a few miles ahead of them a crack had opened in the ocean floor. What they had experienced was a kickback from the rip. It was traveling at two miles a second, and within eleven minutes of that rip, seismographs in Tokyo's Imperial University were busy recording the wave motions that were traversing the world from the rupture along the San Andreas Fault.

A few moments later San Francisco was hit by the "big one." It arrived with a foreshock at 5:12 A.M., and was followed by the great earthquake twenty seconds later. The violent shocks lasted 45 to 60 seconds. This crack in the earth was 290 miles long. The morning of April 18, 1906, would be remembered by generations to come.

The earthquake shock was felt from Coos Bay, Oregon, to Los Angeles and as far east as central Nevada— an area of about 375,000 square miles. Almost immediately all telephone and telegraph communications stopped within the city. The cable at Ocean Beach remained in temporary operation. But postal telegraph operators transmitted their last message to the outside world at 2:20 P.M. as the army ordered them from the building with the fire approaching.

Right after the shaking stopped a messenger arrived at Fort Mason from General Funston. All available troops were to report to Mayor Schmitz at the Hall of Justice at 7:00 A.M.

A little over an hour later there was a major aftershock that caused the collapse of many damaged buildings. All over the city large gray and black smoke appeared as fires began. To make a bad situation worse, the main fire alarms for the city had been destroyed by the quake and no fire alarms sounded. Even if firefighters were able to respond, most of the cities old water pipes cracked open, so there would have been little, if any, water pressure to fight a fire. As the Mission District began to burn, three thousand volunteers and a few firemen fought the blaze with brooms and a little water from a hydrant at 20th and Church.

Meanwhile the army troops began to blast away the debris and piles of brick and wood that laid in the way of the fires. At 8:00 A.M. seven companies of Coast Artillery troops, additional troops from divisions I and K of the 14th Cavalry, plus three batteries of field artillery arrived down town to take up patrol. The army began dynamiting buildings around the U.S. Mint building, in the hope that it could be saved, and it was. The signal corps swung into action and established the Ferry Building telegraph operations a 2;30 P.M.

Mayor Schmitz proclaimed that three men had already been shot without mercy for looting, and let it be understood that the order has been given that all soldiers and police shoot to kill in any cases of miscreants who may seek to take advantage of the city's awful misfortune.

All over the city major fires burned and huge buildings collapsed onto the streets below. St. Mary's hospital at First and Bryant Streets was abandoned as the fire approached it at 1:00 P.M. The fires also destroyed many of the city's landmark buildings and hotels, including the St. Francis at Union Square, The Winchester Hotel, The Hearst

Building at Third and Market Streets, The Mechanic's Pavilion, The Spreckles Building, and all the buildings at the Portsmouth Square.

General Funston sent a telegram to the War Department asking for thousands of tents and all available rations, as people were congregating near and into Golden Gate park and making make shift cabins of wooden planks and using bed sheets for tents. The U.S. government answered General Funston's call for help by sending 200,000 rations to San Francisco. In addition Secretary of War, Howard Taft ordered all hospital, walls, and tents from army posts at Vancouver and from Fort Douglas, Fort Logan, Fort Snelling, Fort Sheridan, and Fort Russell.

Fire fighters attempted to make a stand at 9 P.M. along Powell Street, but they couldn't contain it as it swept up Nob Hill. When the great wall of fire reached Van Ness Avenue at night, the army dynamited mansions along the street, in an attempt to build a firebreak. At the foot of Van Ness Avenue, 16 enlisted men and two officers from the U.S.S. Chicago supervised the rescue of 20,000 refugees fleeing the fires; it was the largest evacuation by sea in history.

Some time between April 18 and 19, Rudolph Speckles and James D. Phelan, the former mayor, entered Mayor Schmitz's bunker office. Their purpose was to arrest Schmitz and Ruef so they could be prosecuted for their malfeasance in office.

Fleeing the fires and burning city, these people are headed to the waterfront. The message reads. "I ain't in that bunch, I stayed downtown all during the fires."

Long lines of people fled to the piers and landings, and to the ships that would take them to Oakland and elsewhere. In the background is the Ferry Building.

The well known Spreckles Building on fire as the long lines of fire hoses lay helpless on the streets. The iron structure of the Spreckles Building resisted the earthquake's force, but the fire that followed gutted it.

In this rare photograph, a daring man watches as a swirling fire seems headed his way on Howard Street. *From the Whelan Collection.*

Despite the dust and soot, these men on Market Street and Front Street don't seem to mind the inferno ahead of them. Because of the lack of water, a useless fire truck sits idly by.

Business men take to the streets, looking stunned at the fire before them on Market Street.

The fire sweeping north on Front Street envelops the Wellman Peck building.

Smoke and fire rage over the buildings on Third Street near Market Street. The dust and debris cover the men pictured in this photograph.

A rare photo taken from the roof top of the Fairmont hotel showing the huge fire raging over the city.
*From the Whelan Collection.*

A view of the "Great San Francisco Fire on the second day," April 19, 1906. Photograph by J.B. Moller.

New explosions rip through an area to the left as the fire burns all day long on the April 19.

A view of the burning city from a safe distance, as photographer Moller records the second day of the fire.

A photo postcard showing people on high ground looking down on the burning city, much like people today view a fire works display show, but this could have been their homes burning. April 18, 1906.

A post card image of the city as the fires are beginning. Many house's in the foreground have not caught fire. The tall building in the center is the Speckles Building. Though enhanced by an artist, this picture stands in stark contrast to the panoramic view of the same area seen on page **XXX**

The long walk back to the shell of a once proud city, as passersby look over what remained of the Palace Hotel.

The Boericke Se Runyon Companies sign stands tall among the ruins.

Someone drove their auto into the badly torn up streets of the city in the hope of finding some one or some thing that had been spared.

Powell Street as seen from Market Street, with a lone street lamp intact.

The army attempted to stop the spread of fire by destroying some buildings in its path to create a fire lane, as can be seen in this photograph. There efforts did little to prevent total destruction in the area.

The shock of the catastrophe has people in a stunned awe, as they walk trancelike through the ruins.

The Conservatory on Strawberry Hill at Golden Gate park, after the earthquake.
Three figures stand among the rubble.

Demolished homes in what was once a large residential neighborhood, captured in this stereo card image.

The resilient human spirit in smiling face and a sidewalk stove outside a ruined home.

The ruined structure of *The Call* buildings sits among spokes of metal fencing.

City Hall at Eighth and Market Streets, looked like this after the quake passed and fire was extinguished.

The wonderful Majestic Theatre was a collapsed shell after the event.

The same fate was the lot of the Tivola Opera House, as captured by photographer Hodson.

Some one left their horse and buggy to look around the empty shells of the buildings.

This stereo photo shows a man who prayed that some things were saved in his locked safe after the fires. He holds a box but what of the contents?

The U.S. Cavalry about to patrol San Francisco under an order of martial law.

A view of the business district with its hollowed out buildings was a dreadful sight, and one worth sending to others on this postcard.

A view of the torn-up city streets after the earthquake.

A sign posted soon after the fire ended reads, "A class 'A' earthquake & fire proof building will be built at once on this property."

The Emporium Building after the quake.

The ruins of the Marchand's Restaurant and the City of Paris Dry Goods store.

A motor car zooms past, between the photographer and the shell of City Hall.

In need of food and other supplies citizens sift through the ruins for canned goods in an area that was a grocery store.

Makeshift housing like this lean-to, made of sticks, boards, and trunks, would serve as housing until something better could be built.

Looking west from the Kohl Building all one can see is devastation. At the top center is the Fairmont Hotel.

A motor car rambles over the treacherous, beaten-up streets.

Carrying what ever they could salvage, these folks stop to look at the photographer. The home on the right is about to collapse.

The Red Cross and the U.S. Army set up this food camp for anyone who came.
Rich or poor, tramps, and the homeless all shared in the meals.

Bread lines set up by the Red Cross sometimes grew to be blocks long.

At a tent camp set up by the police, signs were posted to help people know where to get help, food, and water.

Slowly people made their way back into the devastated city, to survey the damage, see what could be salvaged, and begin to pick up the pieces of their lives.

The Examiner Newspaper Building after the quake and fire.

The ruins of the Flood Mansion.

The tower of the St. Dominie's Church was destroyed by the quake, leaving only the infrastructure and the top of the dome.

Several onlookers stand at the entrance of City Hall after the earthquake.

The ruins of St. Catherine Cathedral after the fire.

A tent was set up outside of the ruined Hall of Justice building on Kearney Street.

The Ferry Building tower sustained some damage, but its grounds were converted to a tent city for the homeless.

A great birds eye view of East Street at the water front, looking toward Telegraph Hill. The earthquake's destruction is evident, as is the temporary tent city. But new construction is in progress.

WATER FRONT.

The photograph of Jefferson Square shows the many refugees living in tents.

With a sense of humor these folks grin and bear it while living in their makeshift houses and tents. Some of the signs read "Ring the bell for Landlady," "The House of Mirth," "No Peddlers," "The Whole Dam Family," and "Camp 346."

The top of this post card shows the burnt out section of San Francisco after the earthquake and fires. The bottom view is the same area, three years later, showing the vast progress being made toward recovery.

Two men observe the shattered homes in the aftermath of the quake. *From the B Whelan Collection.*

Only the shells of buildings remain in this photo of downtown San Francisco. *From the B Whelan Collection.*

This view of the Ferry Building from the water may have been taken before the quake, as we have been told that the top of the building was leaning after the quake. *From the B Whelan Collection.*

After the quake street cars were converted into housing for the homeless. *From the B Whelan Collection.*

Near City Hall, tent cities were erected as temporary shelter. *From the B Whelan Collection.*

# Rebuilding Hope: The Aftermath

Almost as soon as it ended, San Francisco's population began to rebuild the city. Even places like Chinatown, which had been virtually destroyed, began to dig out of the ashes and rebuild the neighborhood. On April 23, 1906, Governor Pardee of California told a newspaper reporter that the work of rebuilding San Francisco had begun, and that he expected "to see this great metropolis be replaced by a much larger scale then ever before." Many countries around the world offered relief and help, but President Theodore Roosevelt declined the offers.

The earthquake and fire left more than two thirds of San Francisco's population with no homes. Combining relief monies and donations, a fund was established to build 5,600 small cottages for the homeless. They were called "earthquake shacks," and were placed row after row in parks around the city. The city charged a token rent ranging from $1.00 to $2.00 a month, and at the end of the year the city agreed to pay back all the rent they had received if the people would remove the shacks by a certain date.

Months after all danger had passed, the refugees, who had made their homes in the parks, refused to leave. They had made new friends and a new life. Even though there were few amenities, and their main sources of food were bread lines and beans. Some went to their old neighborhoods, but found that where their homes had been there remained only piles of dust and rubble.

In the spring of 1909, a street naming commission was appointed by mayor Taylor. The old San Francisco grew up in such a way that various neighborhoods had used popular street names, often distinguished only by the type by adding avenue, street, or boulevard to the name, and sometimes not even that. There were, for example four thoroughfares named Church and Virginia before the earthquake. To complicate matters even more, a street would change its name after a few blocks. This created much confusion among the populace, a minor annoyance at least, and a major safety issue in times of emergency. Taking advantage of the revolutionary changes caused by the earthquake, the commission decided to address the names of streets. The new streets used names of historical or patriotic significance, but none named for a living person.

Great energies went into the recovery of San Francisco, and the result was a new and livable city. The crowning touch and symbol of the rebirth of the city was to be a world's fair. In 1911, President Howard Taft declared that San Francisco would host the Panama Pacific Exposition in 1915.

Work began almost at once. Grand buildings were designed to show off the modern industries and arts, not only of San Francisco, but of the whole world. Interestingly, the buildings were made of plaster materials designed to last only a few months, and much of the site has disappeared. But the Palace of Fine Arts was not torn down and was completely reconstructed in the 1960s.

The fair itself ran from February 20 until December 4, 1915. It was a great success and marked the return of a world-class city.

New construction began almost immediately, even before all the rubble has been removed

At Union Square, the St. Francis Annex and office buildings are in the course of being built again. In the background is the Spreckles Building.

Fillmore Street was again a bustling area after the terrible fires were put out.

A lone sentry guarding the gates to the Presidio army base.

The Ferry Building's tower looms high over the traffic on Market Street.

San Francisco's Market Street once again was the busy hub of the city.

Fillmore street cars and populace crowd one another, in this photo that could have been taken by the Logeman studio, seen in the background.

The Spreckles building was repaired and put back into service, and other large stores were once again doing business as usual.

A photograph of Market Street near Kearney Street, bears witness to the reinvigorated city.

A sure sign of recovery was the reopening of neighborhood stores and businesses. It is also interesting to note the billboard for a March 2 concert by Harold Bauer with the San Francisco Symphony. The Symphony's first concert following the earthquake was in December 1911.

A nice portrait of Mr. And Mrs. Wong Sun Yuw, in their newly stocked store.

Despite the presence of debris, even the fortune tellers were back on the street conducting their business.

Fisherman's wharf was again a scene of bumping boats and casting off. This photo postcard had the title, "A bit of Naples."

This lineup of automobiles outside of the Russell House seems do draw a group of admirers. Or perhaps it the Anheuser-Busch Pilsner they are serving inside.

Auto racing proved to be a great sport for San Franciscans, as this photo shows us. Including a sign posting that Miss Edith Wilson of the city, will appear for today only. Edith Wilson was a race car driver of some renown in early 20th century California.

A Shrine Parade in downtown San Francisco goes on regardless of the rain.

The Ferry Building and adjoining structures surrounded by automobiles.

Another view of the Ferry Building. This center of transportation brought a constant flow of trolleys, automobiles, and wagons.

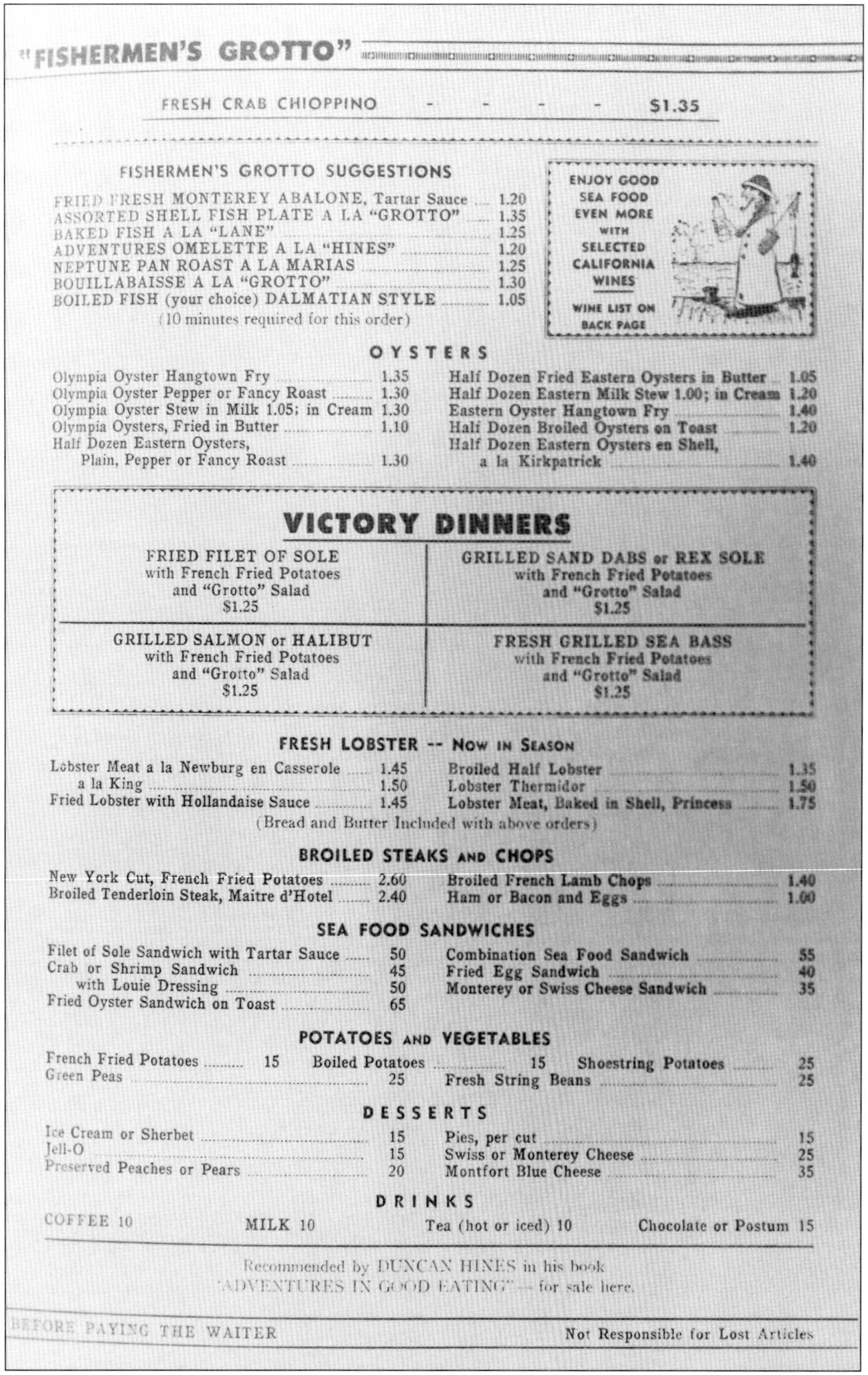

"FISHERMEN'S GROTTO"

FRESH CRAB CHIOPPINO - - - - $1.35

**FISHERMEN'S GROTTO SUGGESTIONS**

| | |
|---|---|
| FRIED FRESH MONTEREY ABALONE, Tartar Sauce | 1.20 |
| ASSORTED SHELL FISH PLATE A LA "GROTTO" | 1.35 |
| BAKED FISH A LA "LANE" | 1.25 |
| ADVENTURES OMELETTE A LA "HINES" | 1.20 |
| NEPTUNE PAN ROAST A LA MARIAS | 1.25 |
| BOUILLABAISSE A LA "GROTTO" | 1.30 |
| BOILED FISH (your choice) DALMATIAN STYLE | 1.05 |

(10 minutes required for this order)

ENJOY GOOD SEA FOOD EVEN MORE WITH SELECTED CALIFORNIA WINES

WINE LIST ON BACK PAGE

**OYSTERS**

| | | | |
|---|---|---|---|
| Olympia Oyster Hangtown Fry | 1.35 | Half Dozen Fried Eastern Oysters in Butter | 1.05 |
| Olympia Oyster Pepper or Fancy Roast | 1.30 | Half Dozen Eastern Milk Stew 1.00; in Cream | 1.20 |
| Olympia Oyster Stew in Milk 1.05; in Cream | 1.30 | Eastern Oyster Hangtown Fry | 1.40 |
| Olympia Oysters, Fried in Butter | 1.10 | Half Dozen Broiled Oysters on Toast | 1.20 |
| Half Dozen Eastern Oysters, Plain, Pepper or Fancy Roast | 1.30 | Half Dozen Eastern Oysters en Shell, a la Kirkpatrick | 1.40 |

**VICTORY DINNERS**

| | |
|---|---|
| FRIED FILET OF SOLE with French Fried Potatoes and "Grotto" Salad $1.25 | GRILLED SAND DABS or REX SOLE with French Fried Potatoes and "Grotto" Salad $1.25 |
| GRILLED SALMON or HALIBUT with French Fried Potatoes and "Grotto" Salad $1.25 | FRESH GRILLED SEA BASS with French Fried Potatoes and "Grotto" Salad $1.25 |

**FRESH LOBSTER -- Now in Season**

| | | | |
|---|---|---|---|
| Lobster Meat a la Newburg en Casserole | 1.45 | Broiled Half Lobster | 1.35 |
| a la King | 1.50 | Lobster Thermidor | 1.50 |
| Fried Lobster with Hollandaise Sauce | 1.45 | Lobster Meat, Baked in Shell, Princess | 1.75 |

(Bread and Butter Included with above orders)

**BROILED STEAKS AND CHOPS**

| | | | |
|---|---|---|---|
| New York Cut, French Fried Potatoes | 2.60 | Broiled French Lamb Chops | 1.40 |
| Broiled Tenderloin Steak, Maitre d'Hotel | 2.40 | Ham or Bacon and Eggs | 1.00 |

**SEA FOOD SANDWICHES**

| | | | |
|---|---|---|---|
| Filet of Sole Sandwich with Tartar Sauce | 50 | Combination Sea Food Sandwich | 55 |
| Crab or Shrimp Sandwich | 45 | Fried Egg Sandwich | 40 |
| with Louie Dressing | 50 | Monterey or Swiss Cheese Sandwich | 35 |
| Fried Oyster Sandwich on Toast | 65 | | |

**POTATOES AND VEGETABLES**

French Fried Potatoes 15 Boiled Potatoes 15 Shoestring Potatoes 25

Green Peas 25 Fresh String Beans 25

**DESSERTS**

| | | | |
|---|---|---|---|
| Ice Cream or Sherbet | 15 | Pies, per cut | 15 |
| Jell-O | 15 | Swiss or Monterey Cheese | 25 |
| Preserved Peaches or Pears | 20 | Montfort Blue Cheese | 35 |

**DRINKS**

COFFEE 10 MILK 10 Tea (hot or iced) 10 Chocolate or Postum 15

Recommended by DUNCAN HINES in his book "ADVENTURES IN GOOD EATING" — for sale here.

BEFORE PAYING THE WAITER Not Responsible for Lost Articles

The Town Taxicab Company was located next door to the Chevrolet dealer. That's a taxi on the left and brand new Chevrolets on the right, lined up for inspection.

This related photograph gives a better view of the Golden Gate Auto Security Company, the Chevrolet dealer. They were offering a great deal of $60 off on a new Chevrolet 490 was bought by January 15th. The 490 was introduced in 1916 to compete with Fords tin Lizzie. Interestingly, $60 was the cost of electric lights and starter set. The car got its name from its original selling price of $490, and was manufactured until 1922.

A panoramic photograph showing the Mt. Tamalpais, California, Railroad, a sight-seeing train. It was pulled by Shay Geared steam engine, normally used in the logging camps of the Pacific Northwest. Both passengers and crew have stepped out of the cars to pose for the photographer. The scene is above the Golden Gate of San Francisco and the line was called the "Crookedest railroad in the world."

MT TAMALPAIS
MT TAMALPAIS
AND
MUIR WOODS
8

The Harbor Entrance to the Pan Pacific Exposition as seen through the lens of the "Official Photographers" of the Expo, the Cardinell-Vincent Company.

The Pan Pacific Expo. Photographed at night.

Another night display of lights at the Pan Pacific Expo.

A panoramic photograph showing a lineup of historic Cadillacs, ranging from the 1902 Model B-1 to the latest model from 1916. Included is the "First 4-cylinder Cadillac ever built," with a sign reading "I am TEN years OLD and still working."

A great panoramic photo of the start of the 'Grand Prix Race" at the Pan Pacific Expo. In Feb 27, 1915. Among the race car drivers was Barney Oldfield. From the signs on the rail it looks like the car of choice was the Maxwell, but white Stutz is in the first row of the lineup.

A panoramic photograph of the Yacht Harbor on Preservation Day, Feb 20, 1916. A Cardinell-Vincent Company photograph.

In Yacht Harbor at the Expo, replicas of early Spanish ships are displayed.

Another view of replicas of Spanish ships at the Pan Pacific Expo. Here it is clear how the Spanish ship were constructed atop modern hulls.

The Palace of Horticulture at the Pan Pacific International Exposition.

The Tower of Jewels.
This tower was lighted
by hundreds of bulbs
when night fell
on the fair.

Details of one of the massive doorways at the Exposition. A Cardinell-Vincent Company photograph.

This large pool ran in front of several buildings, including the Tower of Jewels.

One of the many statues that were around the Exposition, this one caught the eye of an unknown photographer.

A marvelous fountain at the Pan Pacific Expo that shot water fifty feet into the air.

A dramatic night photograph of a pool and the Tower of Jewels. A Cardinell-Vincent Company photograph.

Warships of the Navy Fleet are seen in this night photograph showing off in the San Francisco Bay.

In Golden Gate Park these columns were called' Portals of the Past.

The Grand Stand at Golden Gate Park with people assembling for an event.

The Golden Gate Bridge on the day after it opened in 1937.

The sailing ship "Danmark" passing under the Golden Gate Bridge.

The boats at dock in San Francisco's Fishermen's Wharf.

A rebuilt Cliff House offers a nice image for a postcard photographer in the 1940s.

# Would you like to write a book?

For over thirty years we have had the privilege and pleasure of meeting hundreds of people with wonderful ideas for new books. For many of them, writing a book was only a far off dream, something they hoped to do someday. With our encouragement they went on to produce a beautiful, useful addition to the literature in their field.

If you are one of those people who always dreamed of writing a book, we would love to hear from you. New ideas and authors are the lifeblood of what we do here. We have gathered together a team of people who will do everything they can to make the concept of a book become a reality.

We are eagerly seeking authors to write illustrated, regional history books. With vintage photographs and postcards these big, 11" x 8-1/2" books are bringing bygone eras back to life. Illustrated in glorious full color, each book is stunning & unique.

The following form has helped hundreds to gather their thoughts and begin the process of compiling a book. Submit your idea to us in one of the following ways:

• Email the following information to info@schifferbooks.com

• Copy and mail this form to Acquisitions, Schiffer Publishing, Ltd. 4880 Lower Valley Rd., Atglen, PA 19310

• Fax us at 610-593-2002

We look forward to hearing from you!

Name ______________________

Street Address

______________________

______________________

______________________

Home Phone ______________________

Work Phone ______________________

Fax ______________________

Email address: ______________________

Website: ______________________

Best time to call: ______________________

The subject of the book you are proposing: ________

______________________

______________________

______________________

______________________

Your background in the field: ______________________

______________________

______________________

______________________

______________________

______________________

The number of postcards/historic images you own or have access to: ______________________

A description of your target audience, i.e. population, relevant organizations, etc.: ______________________

______________________

______________________

______________________

______________________

______________________

Specific retailers who would carry your book:

______________________

______________________

______________________

______________________

______________________

______________________

______________________

______________________

______________________

______________________

Media that would review/promote your book:

______________________

______________________

______________________

______________________

______________________

______________________

**Schiffer Publishing, Ltd.**
**4880 Lower Valley Road**
**Atglen, PA 19310**
**610-593-1777**
**www.schifferbooks.com**